Maths
The Basic Skills

Measures
Shape and
Space

Veronica Thomas

Series contributors
June Haighton
Deborah Holder

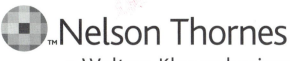

™ Nelson Thornes

a Wolters Kluwer business

Published in 2006 by:
Nelson Thornes Ltd
Delta Place
27 Bath Road
CHELTENHAM
GL53 7TH
United Kingdom

06 07 08 09/10 9 8 7 6 5 4 3 2 1

A catalogue record for this book is available from the British Library

ISBN 0 7487 8330 X

Cover photograph by Don Farrall / Photodisc 85 (NT)

Illustrations by Tech-Set Ltd, Gateshead, Tyne & Wear
Page make-up by Tech-Set Ltd, Gateshead, Tyne & Wear

Printed and bound in Croatia by Zrinski

Contents

Money

Coins

Calculating with money

Money skill check

Time

Days and dates

Tell the time

Time skill check

Measures

Length

Weight

Shape and space

Measures, shape and space mock tests

Answers

Choose coins

5p   £1   £2   20p

2p   50p   10p   1p

Which coins do you need?

1

Cans   50p
Bottles 70p

☐   1 can

2

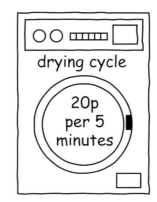

drying cycle

20p
per 5
minutes

☐   5 minutes

3

£1 for trolley

☐   1 trolley

4

Bubblegum

10p for 1
bubblegum

☐   1 bubblegum

Which coin?

Draw a line to the correct answer.

1 Which coin is worth 20p?

2 Which coin is worth £1?

3 Which coin is worth 10p?

4 Which coin is worth 2p?

5 Which coin is worth 5p?

6 Which coin is worth 50p?

7 Which coin is worth the most?

8 Which coin is worth the least?

9 Write the coins in order of value.

Worth least   **Worth most**

1p _____   _____   _____   _____   _____   _____   _____

Which coins or notes do you need?

1 One wash

2 Weekly bus pass

3 One go

4 One card

5 Put the coins and notes in order with the largest value first.

Worth most   **Worth least**

_____   _____   _____   _____

You need a selection of coins: 1p, 2p, 5p.

Select the coins shown in each box.

Write the total of the coins in the circle.

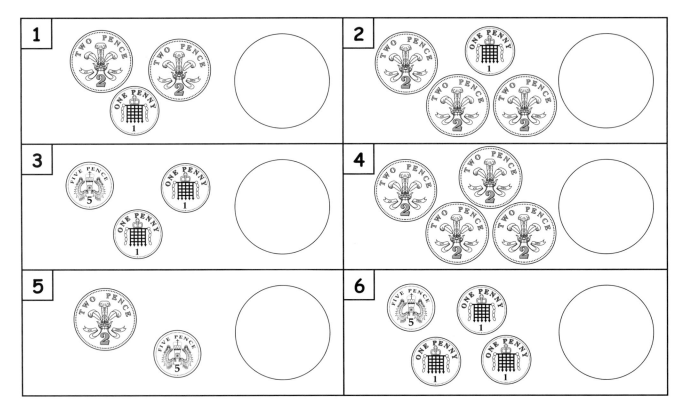

7 There is in a bag. How much is this? _____

8 These coins are in a bag.

How much is this? _____

9 These coins are in the road.

How much is this? _____

10 These coins are on the floor.

How much is this? _____

1 Draw a line to match each bag with the value.

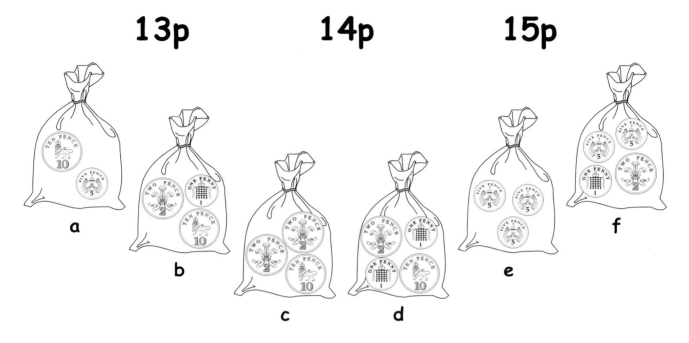

13p **14p** **15p**

a   b   c   d   e   f

2 Tick the boxes that make 20p.

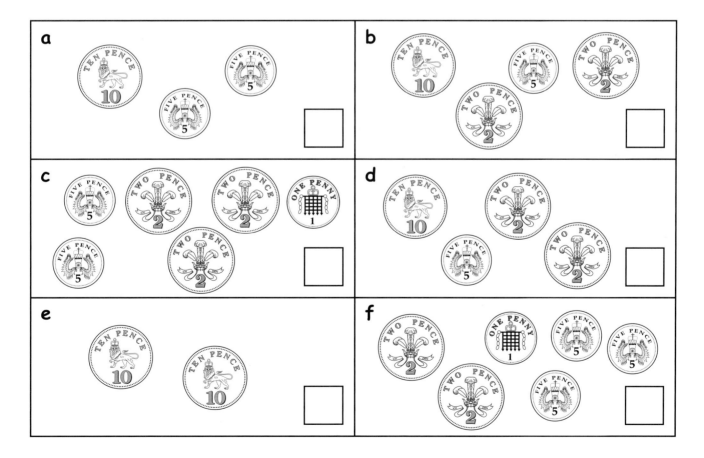

1 Tick the boxes that make 50p.

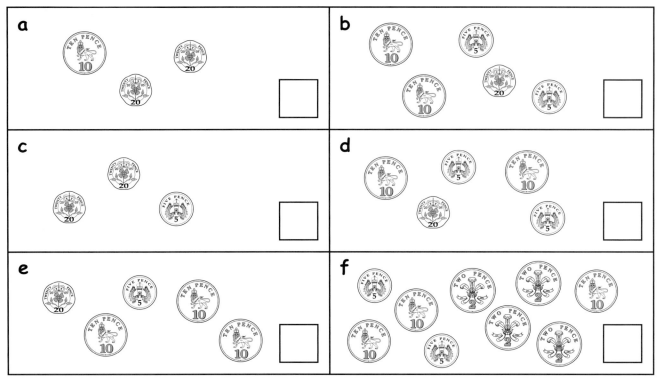

2 Each box of coins has the same value as a single coin. Find the coin. The first one has been done for you.

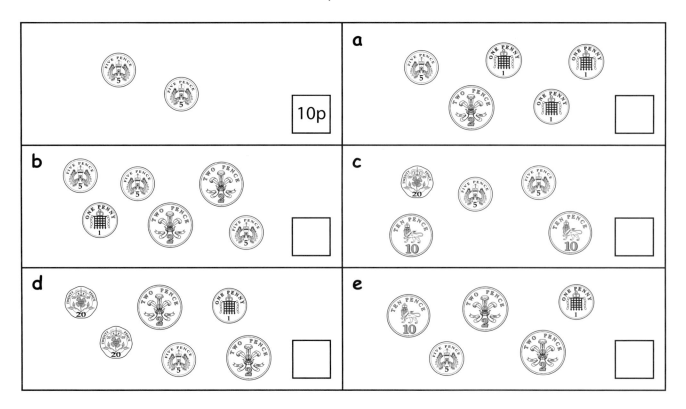

1   Draw lines to match the bags to their value.

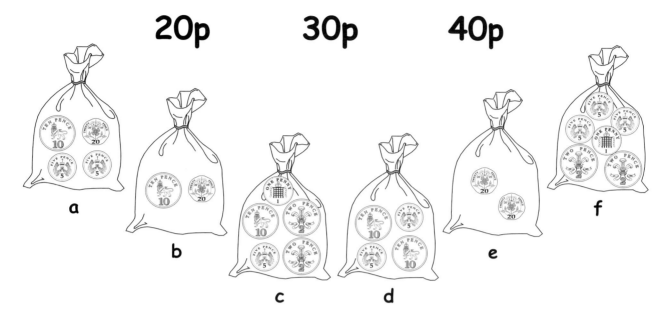

2   Tick the boxes that make £1.

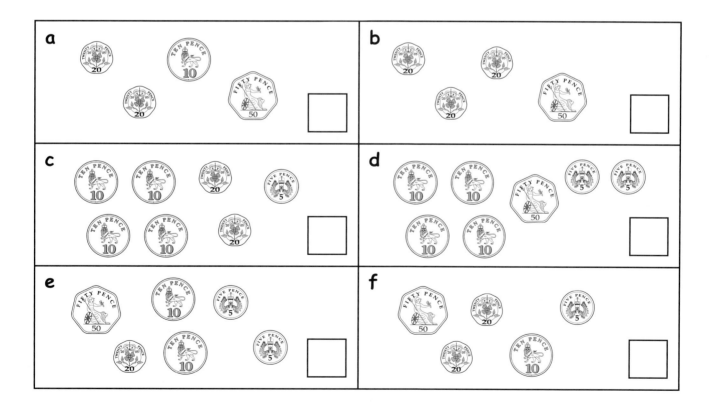

1 Choose 3 coins to pay for each item.
   The first one has been done for you.

tomatoes 23p **20p**   **2p**   **1p**   **a** peas   27p ___ ___ ___

b coffee 65p ___ ___ ___   **c** apple   13p ___ ___ ___

d beans 17p ___ ___ ___   **e** cake   80p ___ ___ ___

f cress 14p ___ ___ ___   **g** biscuits 75p ___ ___ ___

2 Choose 2 extra coins to make the value in each box.
   The first one has been done for you.

28p To make 28p add 20p + 5p	**a** **18p** To make 18p add _____	
b **37p**	**c** **59p**	
d **33p**	**e** **45p**	

Select the coins for each item.

 15p

 12p

Add the coins together.   = 27p

What is the change from £1?

Work out how much you need to add to make £1.

change = 73p

1	bread	52p
	beans	14p
	apple	12p
	total	_____
	change from £1	_____

2	coke	50p
	banana	22p
	onions	17p
	total	_____
	change from £1	_____

3	soup	35p
	peas	30p
	biscuits	32p
	total	_____
	change from £1	_____

4	custard	41p
	beans	14p
	banana	22p
	total	_____
	change from £1	_____

5	pop	20p
	banana	22p
	apple	12p
	total	_____
	change from £1	_____

6	milk	35p
	peas	30p
	beans	14p
	total	_____
	change from £1	_____

7	biscuits	32p
	beans	14p
	cake	31p
	total	_____
	change from £1	_____

8	pop	20p
	cake	31p
	melon	45p
	total	_____
	change from £1	_____

9	jelly	24p
	pears	52p
	apple	12p
	total	_____
	change from £1	_____

Shopping problems

1 Two apples cost 12p each. You give the shopkeeper 50p.

   How much change will you get? _____

2 A peach costs 24p and a banana costs 22p. You give the shopkeeper 50p.

   How much change will you get? _____

3 You buy two bananas for 22p each.

   What change is there from 50p? _____

4 A bar of chocolate costs 42p and a kiwi fruit costs 23p.

   What change will you have from 70p? _____

5 A cucumber costs 55p and cress costs 14p.

   How much change is there from 70p? _____

6 Two peaches cost 24p each and carrots cost 40p.

   a How much is this altogether? _____

   b What change is there from £1? _____

7 Tomatoes cost 45p, lettuce 40p and cress 14p.

   a How much is this altogether? _____

   b What is the change from £1? _____

8 Two peaches cost 24p each and two oranges cost 10p each.

   a How much is this altogether? _____

   b What is the change from £1? _____

9 Two melons cost 32p each and two oranges cost 10p each.

   a How much is this altogether? _____

   b What is the change from £1? _____

Money

You need a selection of coins.

You buy these items. Select the correct coins for each item.

 Peas 27p    rice 65p

Find the total cost. Exchange some coins for a 10p if you can.

Add the coins together.   **92p**

What change is there from £1?   **8p**

Exchange some coins for a 10p to help you find the total.

1 sweetcorn 39p pears 32p total ____ change from £1 ____	**2** soup 48p bread 45p total ____ change from £1 ____	**3** oranges 25p peas 27p total ____ change from £1 ____
4 beans 17p milk 39p total ____ change from £1 ____	**5** tomatoes 24p soup 48p total ____ change from £1 ____	**6** carrots 37p cabbage 36p total ____ change from £1 ____
7 bananas 47p beans 17p peas 27p total ____ change from £1 ____	**8** tuna 53p oranges 25p onions 16p total ____ change from £1 ____	**9** beans 17p peas 27p pears 32p total ____ change from £1 ____

1 A pear costs 15p and a banana 18p.

How much change is there from 50p? _____

2 A cucumber costs 56p and tomatoes 38p.

How much change is there from £1? _____

3 An orange costs 24p and a banana costs 18p.

How much change is there from 50p? _____

4 A cabbage costs 45p and carrots cost 36p.

How much change is there from £1? _____

5 Two peaches cost 24p each and a banana costs 18p.

   a How much is this altogether? _____

   b What is the change from £1? _____

6 Two bananas cost 18p each.

   a How much is this? _____

   b How much change is there from 50p? _____

7 Mushrooms cost 28p, lettuce 45p and cress 14p.

   a How much is this? _____

   b How much change is there from £1? _____

8 Potatoes cost 45p, tomatoes 27p and beetroot 24p.

   a How much is this? _____

   b What change is there from £1? _____

9 Onions cost 17p, garlic 14p and a pepper 32p.

   a How much is this in total? _____

   b How much change is there from £1? _____

Calculate with whole £s

1 Find the cost of each shopping list.

Find the change from £20.

Mark	
chocolates	£5
wine	£7
flowers	£4
total	_____
change from £20	_____

Paul	
nut selection	£8
chocolates	£4
biscuits	£3
total	_____
change from £20	_____

Soumia	
pen set	£7
paints	£6
paper	£4
total	_____
change from £20	_____

2 It costs £3 to get the train to the city, £4 for a cinema ticket and £3 for a burger.
How much change is there from £20? _____

3 Two bottles of wine cost £6 each and chocolates cost £3.
How much change is there from £20? _____

4 Two friends go bowling. It costs £5 each for the bowling and £2 each for the shoes.
They have £15. Is there enough money for each friend to have a £1 bag of chips? _____

5 Look at the price list. Fill in the cost of the food.
Find out how much money each person would have left after paying for each meal.

chips £1   burger £2   pizza £4   kebab £3   curry £5

meal	cost	money left over			
		Deb £10	Max £12	Yui £15	Jag £20
kebab and chips					
curry and chips					
2 burgers and 1 chips					
pizza, kebab and chips					

 p1-3

1 Which coins do you need?

a

50p per hour

1 hour ☐

b

£1 per record

1 record ☐

p2-3

2 List the coins above in order of value.

Worth most   **Worth least**

£2 ____ ____ ____ ____ ____ ____ ____ 1p

3 a Which note is worth the most? _____

b Which note is worth the least? _____

4 Which note is needed? _____

FONES 'R' US

EASY TOP UP CARD

£10

1 Add one coin to make the amount needed in each box.

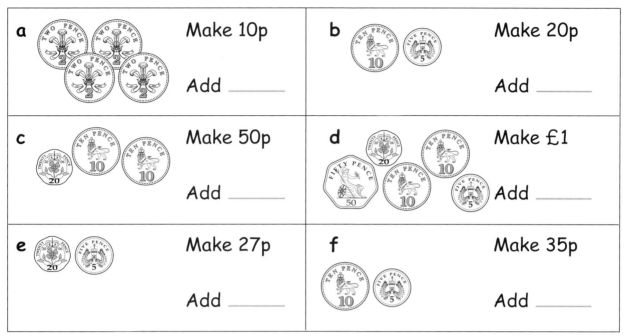

a Make 10p Add _____	**b** Make 20p Add _____
c Make 50p Add _____	**d** Make £1 Add _____
e Make 27p Add _____	**f** Make 35p Add _____

p9-12

2 Add the shopping lists. Find the change.

Pat	
tea	25p
biscuit	22p
total	_____
change from 50p	_____

Jim	
coke	36p
fruit bar	33p
total	_____
change from 70p	_____

Kate	
coffee	45p
cake	34p
total	_____
change from £1	_____

3 Two cokes are 36p each. What is the change from £1? _____

4 A cup of tea is 25p and two cakes are 34p each.

    a How much is this? _____

    b What change is there from £1? _____

p13-14

5 It costs Maroof £4 to go to the gym, £3 for a swim and £2 for a beer. He has £10.
Does he have enough left to pay for his £1 bus fare? _____

6 Two train tickets cost £6 each.
How much change is there from £20? _____

Money skill check 1 answers

1 a   **b**

2 £2, £1, 50p, 20p, 10p, 5p, 2p, 1p

3 a £10   **b** £5

4 £10

Money skill check 2 answers

1 a 2p   **b** 5p   **c** 10p   **d** 5p   **e** 2p   **f** 20p

2 Pat total   47p
        change   3p
     Jim total   69p
         change   1p
     Kate total   79p
         change   21p

3 28p

4 a 93p   **b** 7p

5 yes

6 £8

Money

Tuesday   Sunday   Saturday   Monday

Wednesday   Friday   Thursday

1 Write the days in order starting with Sunday.

Sunday _____   _____   _____   _____

_____   _____

2 Which day is before Wednesday? _____

3 Which day is before Saturday? _____

4 Which day is after Tuesday? _____

5 Which day is after Thursday? _____

6 Which day is after Friday? _____

7 Which days are at the weekend?

_____

8 These are the TV programmes Jenny wants to see.

Monday	EastEnders
Thursday	EastEnders
Wednesday	Coronation Street
Tuesday	Emmerdale
Saturday	Match of the Day

On which days are there no programmes Jenny wants to see?

_____

Seasons

1 Write the season for each picture.

a _____ b _____ c _____ d _____

2 Which season comes after winter? _____

3 Which season comes after summer? _____

4 Which season comes after autumn? _____

5 Which season comes before spring? _____

6 Which season comes before autumn? _____

7 Which season comes before winter? _____

8 Write the seasons in order starting with spring.

   spring _____ _____ _____

There are different ways to write the date.

full date format   11th December 2005

medium date format   11/Dec/2005

| Use the abbreviated month |

short date format   11/12/05

| December is the 12th month | | This means 2005 |

1 Write these dates in the medium date format.

   a 4th November 2005 _____

   b 1st January 2006 _____

   c 23rd March 1995 _____

   d 2nd October 1963 _____

2 Write these dates in the short date format.

   a 21st August 2001 _____   **b** 12th May 2003 _____

   c 3rd November 2005 _____   **d** 11th June 1999 _____

3 Write these dates in full date format.

   a 16/09/04 _____

   b 29/Apr/2005 _____

   c 09/Dec/2002 _____

   d 15/10/05 _____

4 An engineer is called out on the date marked on the calendar.
What is the date? Write it using the short format.

AUGUST 2006

S	M	T	W	T	F	S
		1	2	3	4	5
6	7	8	9	10	11	12
13	14	15	16	17	18	19
20	21	22	23	24	25	26
27	28	29	30	31		

Call out date

Match the dates

Draw lines to match the dates.

7th January 2006	**1**	12/Jun/2006	**a**	7/01/06
13th March 2006	**2**	18/Aug/2006	**b**	1/04/06
18th August 2006	**3**	3/Jul/2006	**c**	13/03/06
12th June 2006	**4**	7/Jan/2006	**d**	21/06/06
1st November 2006	**5**	21/Jun/2006	**e**	22/10/06
30th September 2006	**6**	1/Nov/2006	**f**	3/07/06
1st April 2006	**7**	13/Mar/2006	**g**	2/12/06
2nd December 2006	**8**	1/Apr/2006	**h**	14/05/06
3rd July 2006	**9**	2/Dec/2006	**i**	5/02/06
5th February 2006	**10**	31/Jan/2006	**j**	30/09/06
21st June 2006	**11**	5/Feb/2006	**k**	12/06/06
14th May 2006	**12**	22/Oct/2006	**l**	1/11/06
22nd October 2006	**13**	14/May/2006	**m**	18/08/06
31st January 2006	**14**	30/Sep/2006	**n**	31/01/06

Match the times

Remember

The little hand tells the hour.
The big hand tells the minutes.
If the big hand is on the 12, it means **o'clock**.

2 o'clock

Draw a line to match the times.

1 　　**2** 　　**3** 　　**4**

a 5 o'clock　　**b** 4 o'clock　　**c** 6 o'clock　　**d** 9 o'clock

5 What is the time?

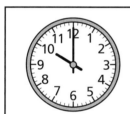　　　　　　

a _____　**b** _____　**c** _____　**d** _____

6 Draw hands to show the times.

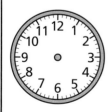　　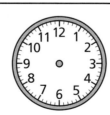　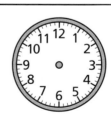　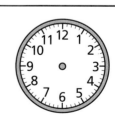

a 2 o'clock　**b** 6 o'clock　**c** 4 o'clock　**d** 8 o'clock　**e** 9 o'clock

Vet's list

1 The vet has a list of all the dogs he has to see.

Write the names of the dogs in the right place on the vet's weekly plan on page 23.

Sam	Monday morning 10 o'clock
Rex	Tuesday afternoon 2 o'clock
Sandy	Monday evening 7 o'clock
Rover	Tuesday morning 10 o'clock
Milo	Friday morning 11 o'clock
Jodie	Thursday afternoon 2 o'clock
Pippa	Thursday afternoon 3 o'clock
Molly	Monday evening 6 o'clock
Daisy	Tuesday afternoon 3 o'clock
Ben	Saturday morning 10 o'clock
Gypsy	Tuesday afternoon 1 o'clock
Bob	Monday afternoon 2 o'clock

2 On which days does the vet see no dogs?

_____

3 On which afternoon does the vet see most dogs?

_____

Time

days	morning am	12 o'clock midday	afternoon pm	evening pm	12 o'clock midnight
weekend Sunday					
Monday					
Tuesday					
Wednesday					
Thursday					
Friday					
weekend Saturday					

All these clocks say **2 o'clock**.

analogue clocks   digital clock

hours   minutes

big hand   big hand   big hand   **00** minutes
on 12   on XII   **where 12**
          would be

1 Write the time under each clock.

a _____ **b** _____ **c** _____ **d** _____

e _____ **f** _____ **g** _____ **h** _____

2 Draw the hands on the clocks to show the times.

a 12 o'clock   **b** 5 o'clock   **c** 8 o'clock   **d** 6 o'clock

3 Write the numbers on the digital clocks to show the times.

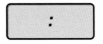 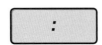  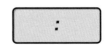

a 12 o'clock   **b** 10 o'clock   **c** 7 o'clock   **d** 1 o'clock

Half hours

All these clocks say **half past 2** or **2 thirty** or **2:30**.

analogue clocks   digital clock

big hand   big hand   big hand   30 minutes
on 6   on VI   where 6
                                 would be

1 Write the time under each clock in all three ways.
   The first one has been done for you.

half past 3   a _____   b _____
3 thirty   _____   _____
3:30   _____   _____

c _____   d _____   e _____
  _____   _____   _____
  _____   _____   _____

2 Draw the hands on the clocks to show the times.

a 5 thirty   b half past 8   c 12:30   d half past 4

3 Write the numbers on the digital clocks to show the times.

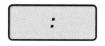   :   :   :   :

a 9 thirty   b 11 thirty   c half past 3   d half past 7

Match the times again

Draw lines between the times that match.

1    half past one   `03:30`

2    half past eight   `09:00`

3    six o'clock   `12:30`

4    half past three   `01:30`

5    twelve thirty   `08:30`

6    ten thirty   `06:00`

7    nine thirty   `09:30`

8   nine o'clock   `10:30`

All these clocks say **quarter past 2** or **2 fifteen** or **2:15**.

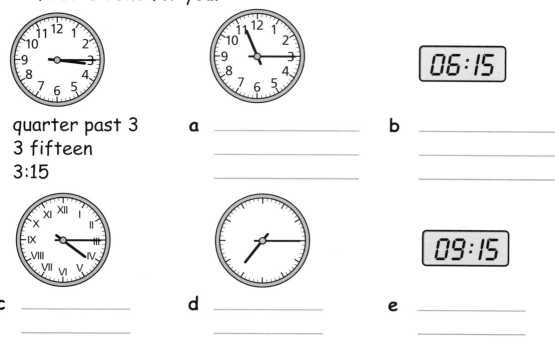

analogue clocks   digital clock

big hand   big hand   big hand   15 minutes
on 3   **on III**   **where 3**
                                    would be

1 Write the time under each clock in all three ways.
The first is done for you.

quarter past 3   **a** _____   **b** _____
3 fifteen   _____   _____
3:15   _____   _____

c _____   **d** _____   **e** _____
   _____   _____   _____
   _____   _____   _____

2 Draw the hands on the clocks to show the times.

a 6:15   **b** seven fifteen   **c** quarter past ten   **d** 2:15

3 Write the numbers on the digital clocks to show the times.

   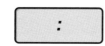      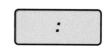

a eight fifteen   **b** quarter past 9   **c** five fifteen   **d** quarter past 1

All these clocks say **forty-five minutes past 2**.
We usually say **quarter to 3**, or **two forty-five**, or **2:45**

analogue clocks   digital clock

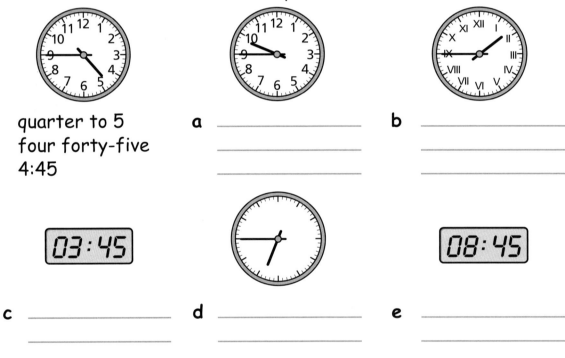

big hand   big hand   big hand   45 minutes
on 9   on IX   where 9
                                    would be

1   Write the time under each clock in all three ways.
    The first one has been done for you.

quarter to 5   a _____   b _____
four forty-five   _____   _____
4:45   _____   _____

c _____   d _____   e _____
   _____   _____   _____
   _____   _____   _____

2   Write the numbers on the digital clocks to show the times.

   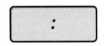   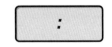   

a quarter to   b quarter to   c two forty-five   d five forty-five
   eight   six

| quarter past |  | quarter to |  |

The minute hand has gone $\frac{1}{4}$ of the way round the clock **past** 12.　　　The minute hand has $\frac{1}{4}$ of the clock to go round to get **to** 12.

1　Write the time under each clock.

a _____　b _____　c _____　d _____

e _____　f _____　g _____　h _____

2　Draw the hands on the clocks to show the times.

a　quarter past 1　b　quarter to 6　c　quarter to 12　d　quarter past 6

e　quarter to 9　　f　quarter to 10　　g　quarter past 2　h　quarter to 1

Match more times

Draw lines between the times that match.

1    quarter to ten   `09:15`

2    quarter past nine   `07:45`

3    quarter to eight   `09:45`

4    quarter past six   `02:15`

5    quarter to nine   `08:45`

6    quarter past twelve   `02:45`

7    quarter past two   `06:15`

8    quarter to three   `12:15`

     Time

1   The library opens at 9:30.

Your watch shows this time.

Is the library open yet?　_____

2   A football match is $1\frac{1}{2}$ hours long.

The match starts at:　 03:00

When does the match end?　_____

3   You go to the pub at:

You stay at the pub for $1\frac{1}{2}$ hours.
When do you leave the pub?
Write the time on a digital watch.　[:]

4   You get home at five thirty.
Your favourite TV programme starts at: 05:45

Are you in time for the start of your TV programme?　_____

5   You are at the bus stop at: 03:45

The bus arrives at half past three. Will you catch the bus?　_____

6   You shop for 3 hours on a Saturday. You start shopping at:

When do you stop shopping?　_____

7   You start cooking dinner at quarter past six.
The dinner needs to cook for $1\frac{1}{2}$ hours.
When will the dinner be ready?
Write the time on the digital clock　 [:]

Time skill check 1   **E1**

p17

1 Write the days in order starting with Monday.

Monday _____   _____   _____

_____   _____   _____

2 Tick the day which is at the weekend.

| Tuesday | Saturday | Thursday |

p18

3 Which season comes before summer? _____

4 In which season is New Year? _____

p21-22, 24

5 Which clocks say 5 o'clock? _____

a 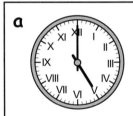 b  c  d

6 Draw the hands on the clock to show 8 o'clock.

7 Tick the sentence which is most likely to be true.

   a Most people wake up in the morning.

   b Most people wake up in the afternoon.

p19-20

1 Write the shortened form of these months.

   a January _____   **b** July _____

   c August _____   **d** November _____

2 Which is the 6th month of the year? _____

3 An engineer is called out on 6th December 2005.

   a Write the date in the box.

Call out date

p25-31

   b The engineer says he will arrive at quarter to 2.

     He arrives at: 02:30

     Fill in the missing word:

late	early	on time

     The engineer is _____ .

   c The engineer stays for $1\frac{1}{2}$ hours.

     At what time does the engineer leave? _____

4 The oven is turned on at five forty-five.

   a Write this time on the digital clock.

   b The dinner has to cook for 30 minutes.

     Draw the hands on the clock to show
     when the dinner is ready.

Skill check

Time skill check 1 answers

1   Monday, Tuesday, Wednesday, Thursday, Friday, Saturday, Sunday

2   Saturday

3   spring

4   winter

5   a and d

6

7   a

Time skill check 2 answers

1   **a** Jan   **b** Jul   **c** Aug   **d** Nov

2   June

3   **a**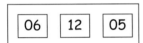

   b late
   c 4 o'clock

4   **a**

   b

Large or small?

1 Which word describes each item?

| large   small |

a _____   b _____   c _____   d _____   e _____   f _____

2 Fill in the missing words.

| larger   smaller |

pan

cooker

pen

book

a The pan is _____ than the cooker.

b The cooker is _____ than the pan.

c The pen is _____ than the book.

d The book is _____ than the pen.

3 Find the smallest and the largest. The first one has been done for you.

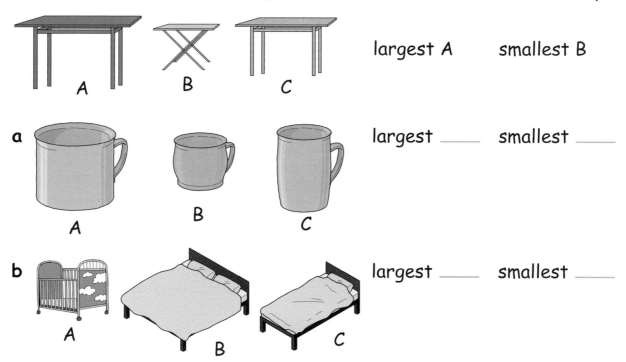

largest A   smallest B

a   largest ____   smallest ____

b   largest ____   smallest ____

Long, wide, tall, short, thin

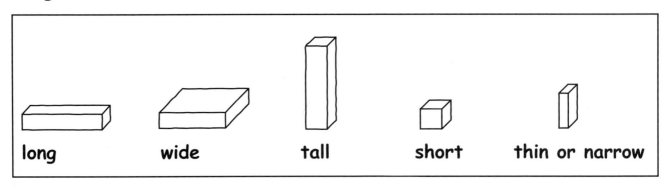

| long | wide | tall | short | thin or narrow |

1 Choose 2 words to describe each dog.

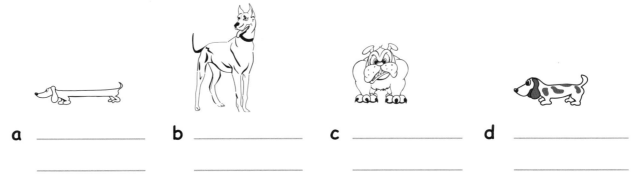

a _____

b _____

c _____

d _____

_____

_____

_____

_____

2 Fill in the missing words.

| longer   shorter   thinner   wider |

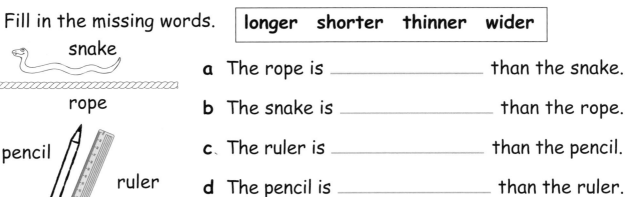

snake

rope

pencil

ruler

a The rope is _____ than the snake.

b The snake is _____ than the rope.

c The ruler is _____ than the pencil.

d The pencil is _____ than the ruler.

3

Ben's hat

 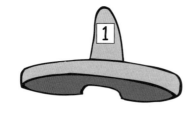 

a Which hats are taller than Ben's hat? _____ _____

b Which hat is wider than Ben's hat? _____

c Which hat is shorter than Ben's hat? _____

d Which hats are narrower than Ben's hat? _____ _____

Measures

Longest, shortest, widest, narrowest, thinnest

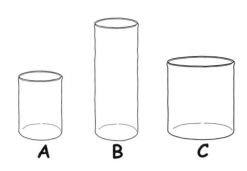

1 a Which glass is the **tallest**?

_____

b Which glass is the **shortest**?

_____

c Which glass is the **widest**?

_____

2 a Which vehicle is the **longest**?

_____

b Which vehicle is the **narrowest**?

_____

c Which vehicle is the **shortest**?

_____

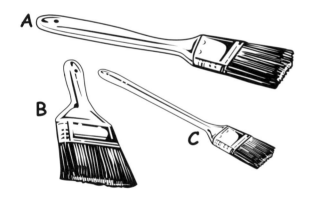

3 a Which brush is the **thinnest**?

_____

b Which brush is the **longest**?

_____

c Which brush is the **shortest**?

_____

4 a Rope A is the _____ .

b Rope B is the _____ .

c Rope C is the _____ .

5 a Tail A is the _____ .

b Tail B is the _____ .

c Tail C is the _____ .

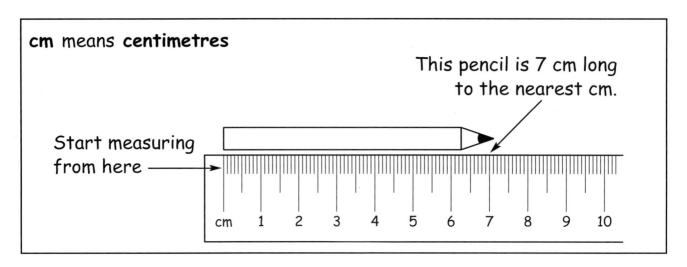

cm means **centimetres**

This pencil is 7 cm long to the nearest cm.

Start measuring from here

cm   1   2   3   4   5   6   7   8   9   10

Measure the length of these lines to the nearest cm.

1 ———————————————————————————— 13 | 14 | 15 |

2 ——————————————

3 —————————

4 ———————————————————

5 ————————

6 ——————————————————————

7 ————————————————————————

8 ——————————————

9 ——————————————

10 ——————————————————————————

11 ————————————————————————————————

12 ——————————————————————————————————

Measure the length of these objects to the nearest cm.

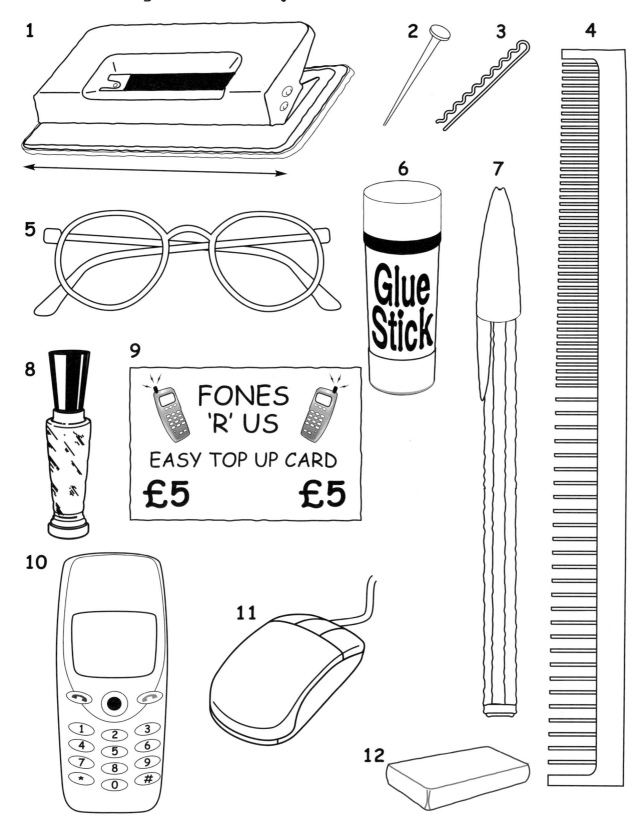

13 Measure this book to the nearest cm.

    a The length of this book is _____ .

    b The width of this book is _____ .

Measure the sides

Measure these shapes to the nearest cm.

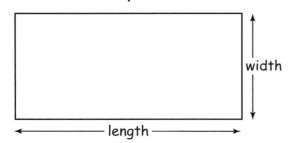

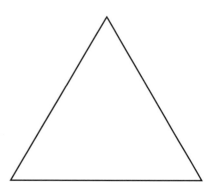

1 a The length of this rectangle is _____ .

   b The width of this rectangle is _____ .

2 The length of each side of this triangle is _____ .

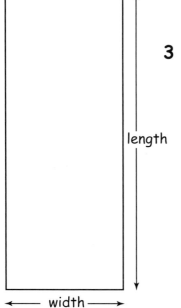

3 a The length of this rectangle is _____ .

   b The width of this rectangle is _____ .

4 The length of each side of this square is _____ .

5 The sides of this triangle measure:

  a _____

  b _____

  c _____

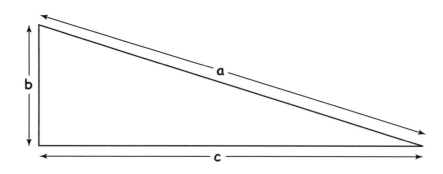

6 a The height of this cylinder is _____ .

   b The width of this cylinder is _____ .

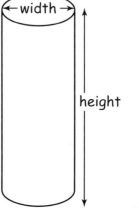

Measures

Estimate and measure

You need a ruler and a selection of objects.

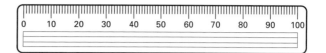

1 Write the name of the first object in the table under **Object**.

2 How long do you think the object is?
   Write your answer in the table under **My estimate**.

3 Check the length with the ruler. Write your answer in the table.

4 Find the difference between your estimate and the actual
   measurement. Write your answer in the table.
   Hint longer measurement take away shorter measurement.

Do the same again using other objects until the table is full.

Object	My estimate	Measured with ruler	Difference (longer – shorter)
This book	28 cm	30 cm	30 cm - 28 cm = 2 cm
	cm	cm	cm
	cm	cm	cm
	cm	cm	cm
	cm	cm	cm
	cm	cm	cm
	cm	cm	cm
	cm	cm	cm
	cm	cm	cm

Metres and centimetres

centimetre (cm)	metre (m)
—— This line is 1 cm long.	100 cm = 1 m

1 Would you use centimetres or metres to measure these items?
Write **m** or **cm**.

a The height of a fridge _____   **b** The length of a bus _____

c The length of a pen _____   **d** The length of a room _____

e The length of a caravan _____   **f** The length of a garage _____

g The height of a chair _____   **h** The height of a glass _____

i The length of a book _____   **j** The length of a sofa _____

2 You must be 1 metre (1 m) or over to
go on rides at an adventure park.
Can these children go on the rides?
Answer **yes** or **no**.

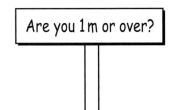

Are you 1 m or over?

a Ben 90 cm tall _____   **b** Tomoko 110 cm tall _____

c April 85 cm tall _____   **d** Jake 100 cm tall _____

e Kristen 120 cm tall _____   **f** Tara 98 cm tall _____

3 Some children measure their height.

a How tall is Jane? _____ cm   **c** How tall is Sam? _____ cm

b How tall is Yusuf? _____ cm   **d** How tall is Amy? _____ cm

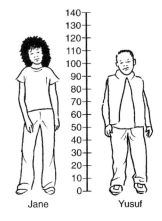

Jane   Yusuf

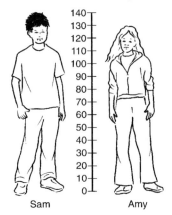

Sam   Amy

Measures

Heavy or light?

1 Which word describes each item?

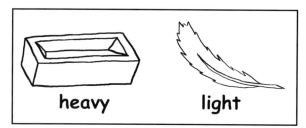

heavy   light

tank   pen   table   £2 coin   paperclip   book   TV

a _____   b _____   c _____   d _____   e _____   f _____   g _____

2 Fill in the missing words.

heavier   lighter

a The tank is _____ than the pen.

b The £2 coin is _____ than the table.

c The TV is _____ than the £2 coin.

d The table is _____ than the tank.

e The TV is _____ than the book.

f The book is _____ than the paperclip.

g The book is _____ than the table.

h The pen is _____ than the paperclip.

i The TV is _____ than the tank.

j The £2 coin is _____ than the book.

k The paperclip is _____ than the table.

Heaviest or lightest?

Example
Which is the heaviest?
Which is the lightest?

heaviest   lightest

Remember The largest is not always the heaviest.

Fill in the missing words.

pan

bread

coin

1 a The **heaviest** is the _____ .

b The **lightest** is the _____ .

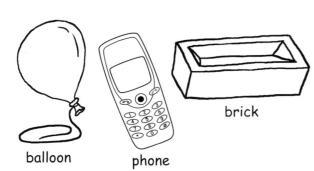

balloon   phone   brick

2 a The **lightest** is the _____ .

b The **heaviest** is the _____ .

feather   milk bottle

computer

3 a The **heaviest** is the _____ .

b The **lightest** is the _____ .

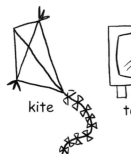  

kite   television   chair

4 a The _____ is the television.

b The _____ is the kite.

Measures

Kilograms or grams?

Grams are used to measure light things.
Kilograms are used to measure heavy things.

A small leaf weighs about 1 g.

1 g

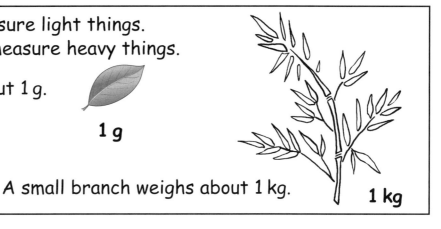

A small branch weighs about 1 kg.

1 kg

1 What would you use to measure these things? Write **kg** or **g**.

a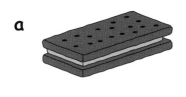

b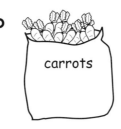

carrots

c

_____

_____

_____

d

e

f

_____

_____

_____

2 Which is more likely? Circle the answer.

a

b

onions

c

CHEESE

d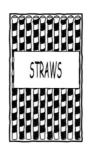

STRAWS

1 g or 1 kg

2 g or 2 kg

100 g or 100 kg

50 g or 50 kg

More or less than 1 kilogram?

1 kilogram (kg) = 1000 grams (g)

A bag of sugar weighs 1 kilogram.

Circle the items that you think are **less than 1 kilogram**.
Hint You can estimate the weight or you can read the label.

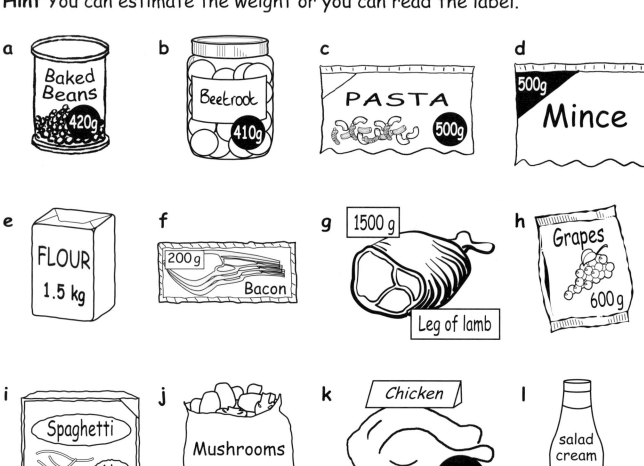

a Baked Beans 420g

b Beetroot 410g

c PASTA 500g

d 500g Mince

e FLOUR 1.5 kg

f 200 g Bacon

g 1500 g Leg of lamb

h Grapes 600 g

i Spaghetti 1 kg

j Mushrooms 100 g

k Chicken 2.4 kg

l salad cream 565 g

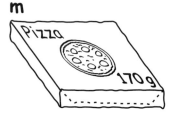

m Pizza 170 g

n Chicken nuggets 275 g

o RICE 1500 g

p New Potatoes 750 g

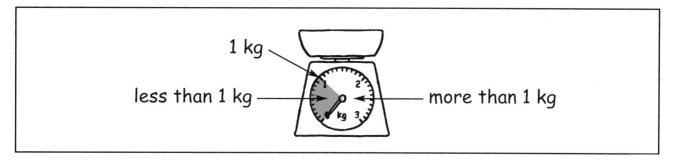

1 kg

less than 1 kg ——————   —————— more than 1 kg

You need kitchen scales and a selection of objects.

1 Choose 2 of the objects. Write them in the table under **1st pair**.

2 Tick the box to show whether you think the 2 objects **together** will be more or less than 1 kg.

3 Find the 1 kg mark on the scales. Check the weight of the 2 objects together on the scales. Tick the correct box.

Do the same again using other objects until the table is full.

Objects	I think these weigh		Checked with the scales	
	less than 1 kg	more than 1 kg	less than 1 kg	more than 1 kg
1st pair				
2nd pair				
3rd pair				
4th pair				
5th pair				

More or less than ½ kilogram?

½ kilogram = 500 grams

1 Tick the items that you think are less than ½ kilogram.

Remember You can estimate the weight or read the label.

a

b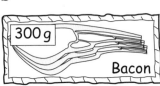

c

d

e

f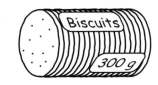

g

h

2 Draw a line between each item and its weight.

a

more than 3 kg

b

2 kg

c

about 1½ kg

d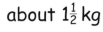

1 kg

e

about ½ kg

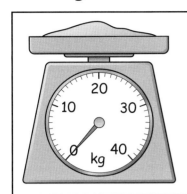

Every 10 kg are labelled.

The lines between show the kg.

There is 1 kg between each mark and the next.

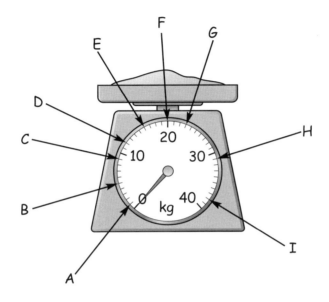

1 Arrow A points to 1 kg.
What do the other arrows point to?

Arrow B points to _____   Arrow C points to _____

Arrow D points to _____   Arrow E points to _____

Arrow F points to _____   Arrow G points to _____

Arrow H points to _____   Arrow I points to _____

2 Put arrows on the scale pointing to these weights.

   a 6 kg   **b** 14 kg   **c** 26 kg   **d** 35 kg

Empty, full, more or less?

Bottle A is **full**.
This has the **most**.

Bottle 3 is **empty**.
This has the **least**.

Bottle B has **less than** bottle A.
It has **more than** bottle C.

1 a Which has **more** drink? _____

b Which has **less** drink? _____

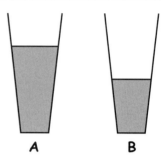

2 a Which has the **most** drink? _____

b Which has the **least** drink? _____

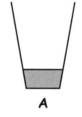  

Choose the right word.

empty	full	more	less

3 a Glass A is _____ .

b Glass B has _____ than glass C.

4 a Mug A is _____ .

b Mug B has _____ than mug C.

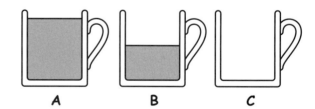

5 a Glass C is _____ .

b Glass B has _____ than glass A.

c Glass B has _____ than glass C.

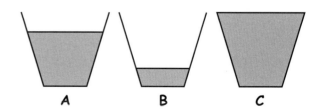

50

Measures

Which holds most?

Remember

If the height is the same
the widest glass holds the most.

The widest glass is glass C.

Glass C holds the most.

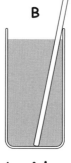

A **narrow**
glass

A **wider**
glass

The **widest**
glass

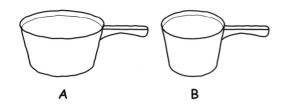

A   B

1 a Which pan holds **more**? _____

b Why? _____

A   B   C

2 a Which bucket holds the **most**? _____

b Why? _____

A   B   C

3 a Which jar holds the **most**? _____

b Why? _____

A   B   C

4 a Which mug holds the **most**? _____

b Why? _____

The amount a container will hold is called its **capacity**.

1 ml = a few drops

Capacity is measured in **litres** (l) and **millilitres** (ml).

Small amounts are measured in millilitres.

Larger amounts are measured in litres.

This bottle holds 1 litre.

What would you use to measure the capacity of these items?
Write **l** or **ml**.

1 ____   2 ____   3 ____   4 ____

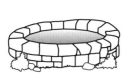  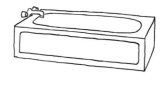 

5 ____   6 ____   7 ____   8 ____

9 Which is more likely? Circle the answer.

a    b    c    d 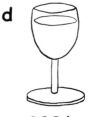   e

a 2 l or 2 ml

b 500 l or 500 ml

c 5 l or 5 ml

d 200 l or 200 ml

e 1 l or 1 ml

1 litre (l) = 1000 millilitres (ml)

1 Tick the items that you think hold more than 1 litre.

Remember You can estimate the capacity or read the label.

a

b

c

d

e

2 This small paint pot holds 1 litre. How many times can it be filled from each of these pots of paint?

a

b

c

d

_____ times   _____ times   _____ times   _____ times

3 You have 2 litres of water.
Tick the containers that will hold the 2 litres.

a

b

c

d

½ litre (l) = 500 millilitres (ml)

1 Tick the items that you think hold less than ½ litre.

a

b

Remember You can estimate the capacity or read the label.

c

d

e

f

g

h

i

2 Draw lines to match each container with its capacity.

a

b

½ litre

c

1 litre

d

1½ litres

2 litres

e

2½ litres

Measure in litres

What capacity do the arrows show?

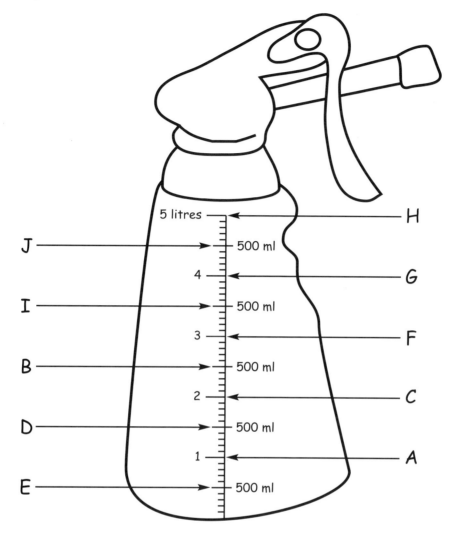

Arrow A shows 1 litre.   Arrow B shows 2 litres 500 ml.

What do the other arrows show?

1 Arrow C shows _____   **2** Arrow D shows _____

3 Arrow E shows _____   **4** Arrow F shows _____

5 Arrow G shows _____   **6** Arrow H shows _____

7 Arrow I shows _____   **8** Arrow J shows _____

Temperature is measured in degrees Celsius and degrees Fahrenheit.
In the UK, we measure temperature in **degrees Celsius** (°C).

30°C	It's a very hot sunny day.
15°C	It's warm.
0°C	It's freezing.

1 Draw a circle around what you might do if the temperature was 30°C.

2 Draw a square around what you might do if the temperature was 15°C.

3 Tick what you might do if the temperature was 0°C.

a

b

c

d

e

f

g

h

i

Warmer or cooler?

Look at the temperatures on the map.

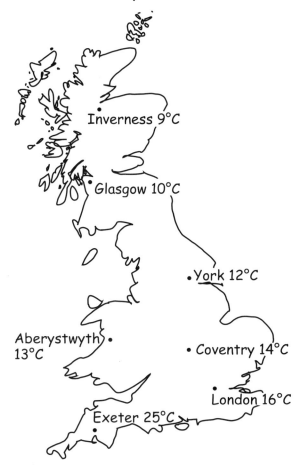

London is 16°C.

Glasgow is 10°C.

London is **warmer** than Glasgow.

Glasgow is **cooler** than London.

Add the word **warmer** or **cooler**.

1   Aberystwyth is _____ than London.

2   London is _____ than Aberystwyth.

3   Aberystwyth is _____ than Glasgow.

4   Glasgow is _____ than Aberystwyth.

5   Exeter is _____ than London.

6   London is _____ than Exeter.

7   York is _____ than Exeter.

8   Exeter is _____ than York.

9   Inverness is _____ than Coventry.

10   Coventry is _____ than Inverness.

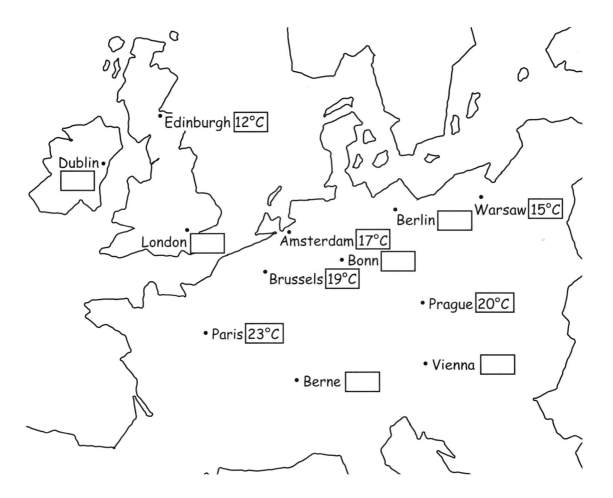

The temperatures are measured to the nearest degree.

Read the sentences below.

Fill in the missing temperatures on the map.

London is **warmer** than Edinburgh, but **colder** than Warsaw.

Bonn is **colder** than Brussels, but **warmer** than Amsterdam.

Berlin is **warmer** than Warsaw, but **colder** than Amsterdam.

Dublin is **warmer** than Edinburgh, but **colder** than London.

Vienna is **colder** than Paris, but **warmer** than Prague.

Berne is **colder** than Paris, but **warmer** than Vienna.

Weather forecasts

Things a weather forecaster might say.

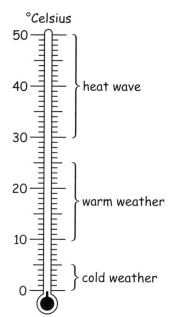

Here is the world weather forecast for
1st December 2005.

city	temperature		city	temperature
Auckland	18°C		Khartoum	37°C
Baghdad	20°C		Kiev	0°C
Bangkok	34°C		London	10°C
Berlin	1°C		Melbourne	16°C
Budapest	2°C		New Orleans	17°C
Calcutta	29°C		Ottawa	0°C
Colombo	30°C		Quinto	15°C
Dakar	33°C		Rome	14°C
Edinburgh	3°C		Singapore	35°C
Hong Kong	25°C		Tokyo	14°C
Islamabad	16°C		Wellington	26°C

1   Which cities around the world had a heat wave on 1/12/2005?

_____

2   Which cities around the world had warm weather on 1/12/2005?

_____

_____

3   Which cities around the world had cold weather on 1/12/2005?

_____

The level of the liquid **rises** as the temperature gets **hotter**.

The level of the liquid **falls** as the temperature gets **colder**.

The liquid shows the temperature in °Celsius.

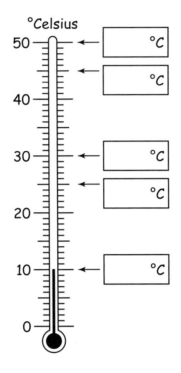

1 Write the temperatures marked by arrows in the boxes.

2 Draw arrows to show these temperatures on the thermometer.
Write the temperature beside each arrow.

5°C   15°C   20°C   35°C   40°C

3 What temperature is the liquid showing on the thermometer? _____

4 Colour the thermometer to show 23°C.

Scales around the house

The fridge is set to number **3**.

 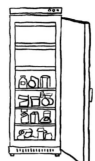

1 These fridge dials are set between numbers.
Write down the **nearest** number setting on each.

  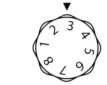

a _____ **b** _____ **c** _____ **d** _____

2 What settings are shown below?

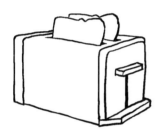 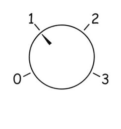 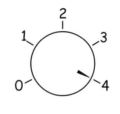 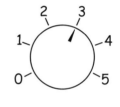

a _____ **b** _____ **c** _____

3 Roughly what is the temperature set at?
Write down the **nearest** number.

 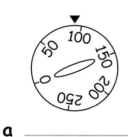 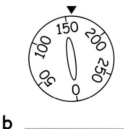 

a _____ **b** _____ **c** _____

4 What temperature settings are shown below?

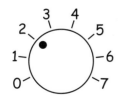 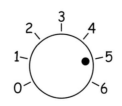 

a _____ **b** _____ **c** _____

Measures

Dials in the car

> **Remember** Ask yourself:
>   What is the **nearest** number?
>   Is the pointer **before** that number?
>   Is the pointer **after** that number?

1 Approximately what is the speed of these cars?

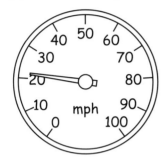 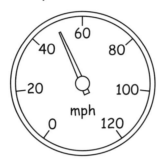 

a _____ b _____ c _____

2 Is the tank **more** or **less** than half full?

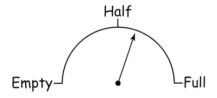 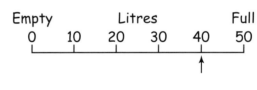 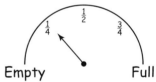

a _____ b _____ c _____

3 Approximately what is the temperature in the car?

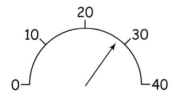 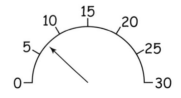 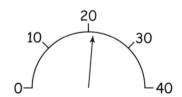

a _____ b _____ c _____

Measures

Fill in the gaps

Look at this scale:

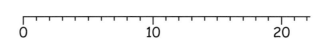

The 0, 10 and 20 are labelled.
There are 9 marers between 0 and 10.
Each marker represents 1.

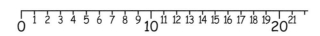

Fill in the missing numbers on these scales:

1

2

Look at this scale:

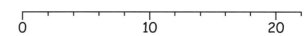

The 0, 10 and 20 are labelled.
There are only 4 markers between
the numbers.
This scale goes up in twos.
Each marker represents 2.

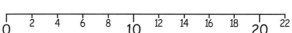

Fill in the missing numbers on these scales:

3

4

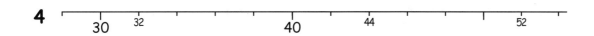

5 What are the measurements shown by the arrows?

a 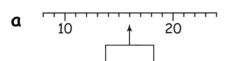   **b**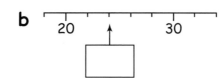

More gaps

Look at this scale.

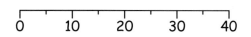

0, 10, 20, 30 and 40 are labelled.

There is one marker between 0 and 10.

This scale goes up in fives.

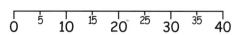

Each marker represents 5.

Fill in the missing numbers on these scales.

1

2
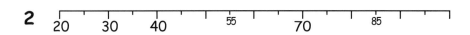

3 Look at this scale.

0, 20, 40, 60, 80, 100 are labelled.

Fill in the missing numbers 10, 30, 50, 70 and 90.

4 Fill in the missing numbers on this scale.

5 What are the measurements shown by the arrows?

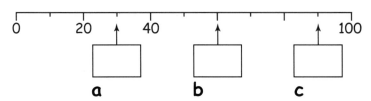

Measures skill check 1

elephant

owl

spider

p35

1 a Which animal is large? _____

   b Which animal is small? _____

   c Which animal is the largest? _____

2 Fill in the missing words.

larger	smaller

   a The owl is _____ than the elephant.

   b The owl is _____ than the spider.

p36-37

3 Fill in the missing words.

> long short thin
>
> taller shorter
>
> longer wide tall

   a The giraffe has a _____ neck.

   b The giraffe has a _____ tail.

   c The monkey is _____ than the giraffe.

   d The giraffe is _____ than the monkey.

   e The monkey's tail is _____ than the giraffe's tail.

Skill check

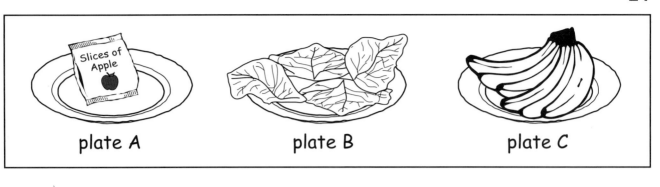

plate A   plate B   plate C

p43-44

4 Fill in the missing words.

| heavy | light | heavier | lighter |

a Plate C is _____ .

b Plate B is _____ .

c Plate A is _____ than plate B.

d Plate A is _____ than plate C.

e Which plate is the heaviest? _____

Here are 3 water bottles.

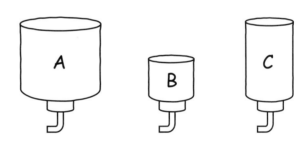

p50-51

5 Fill in the missing words.

| more | less |

a Bottle A holds _____ than bottle C.

b Bottle C holds _____ than bottle B.

c Bottle B holds _____ than bottle A.

d Which bottle holds the most? ____

e Which bottle holds the least? ____

6 Which bottle is full?

_____

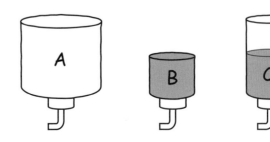

Measures

p38-42

1 Tick the units that measure length.

    cm   **ml**   **kg**   **m**   **g**

2 a Which unit would you use to measure the length of a room? _____

   b Which unit would you use to measure the length of a pencil? _____

3 Estimate the length of these pieces of string.

  a _____   **b** _____

4 Use a ruler to measure the length of this pencil.

                                                    length _____

p42-44

5 How many centimetres are there in 1 metre? _____

6 Children need to be taller than 1 m to go on rides at an adventure park.

   a Jack is 96 cm tall. Can he go on the rides? _____

   b Amal is 105 cm tall. Can she go on the rides? _____

p45-49

7 Draw lines to match the items with the weights.

  a    **b**   **c**   **d**

     1 g   **1 kg**   **5 kg**   **50 g**

Skill check

8 What is kg short for? Circle the answer.

gram   kilometre   kilogram   litre

9 What weight is shown on the scales?

_____

p52-55

10 Which units are used to measure liquids (capacity)?
Circle the answer.

ml   km   l   m   cm

11 What is ml short for? Circle the answer.

millimetres   centimetres   litres   millilitres

12 Draw lines to match the items with the amounts.

a   b    c    d

5 ml   1000 ml   ½ litre   2 l

13 How many times can the teapot be filled from the large urn?

   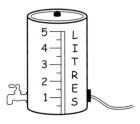

Teapot holds 1 l

The teapot can be filled _____ times.

Measures

p56-60

14 Which units do we use to measure temperature?
Circle the answer.

°Celsius   ml   grams   metres   kg

15 Here is a temperature chart for holiday destinations.

holiday destinations	temperature
Madrid	24°C
New York	15°C
Ottawa	5°C
Paris	20°C
Rome	30°C

a Which holiday destination is the coldest? _____

b Which holiday destination is the warmest? _____

p61-64

16 Fill in the missing numbers on these scales.

a
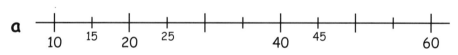

b

17 What is the temperature on the thermometer? _____

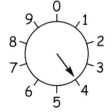

0   10   20   30   40   50
°Celsius

18 What is the temperature on these oven dials?

a
grill

temperature _____

b
oven °C

temperature _____

Measures

Measures skill check 1 answers

1 a elephant   b spider   c elephant

2 a smaller   b larger

3 a long or thin   b short or thin   c shorter   d taller   e longer

4 a heavy   b light   c heavier   d lighter   e C

5 a more   b more   c less   d A   e B

6 a B

Measures skill check 2 answers

1 cm, m

2 a m   b cm

3 a 4, 5 or 6 cm   b 8, 9, 10, 11 or 12 cm

4 14 cm

5 100

6 a no   b yes

7 a 1 kg   b 1 g   c 50 g   d 5 kg

8 kilogram

9 2 kg

10 ml, l

11 millilitres

12 a 1000 ml   b 5 ml   c 2 l   d $\frac{1}{2}$ litre

13 5 times

14 °Celsius

15 a Ottawa   b Rome

16 a

b

17 20°C

18 a 4   b 200°C

Squares

A square has:
4 sides all the same length
4 square corners

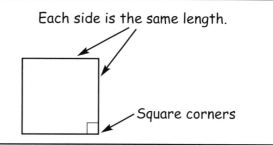

Each side is the same length.

Square corners

1 Tick the shapes that are squares.

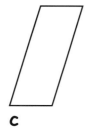

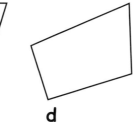

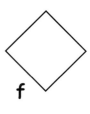

a

b

c

d

e

f

2 How many squares can you see in this pattern. _____

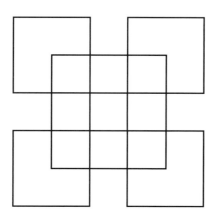

3 Finish drawing these squares.

a

b

c

A rectangle has:
4 sides
2 equal long sides and 2 equal short sides
4 square corners

1 Tick the shapes that are rectangles.

   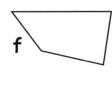

a   b   c   d   e   f

2 Write the name of each shape.

a _____   b _____   c _____   d _____

3 a How many rectangles can you see in this pattern? _____

b How many squares can you see in this pattern? _____

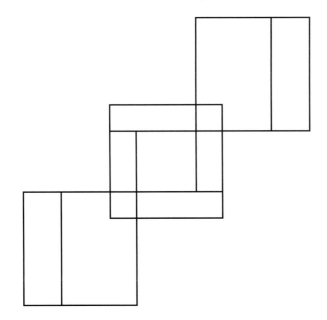

Shape and space

Circles

This is a circle. 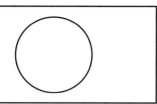

1 Tick the circles.

a   b   c   d   e

2 How many circles are in this pattern? _____

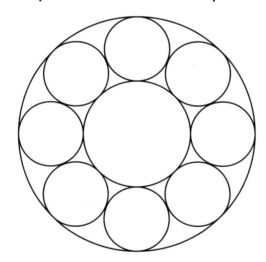

3 Draw in the missing circles to complete the second bicycle.

   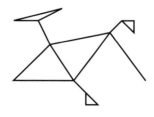

bicycle

Which shape?

1 Draw the next shape in each pattern.
  Write the name of each shape you have drawn.

a
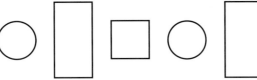

Shape name _____

b
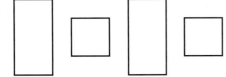

Shape name _____

c

Shape name _____

2 Name the shapes of these signs.

no
cycling

toilets

H

a _____   b _____   c _____

no
smoking

fire
exit

closed

d _____   e _____   f _____

Shape and space

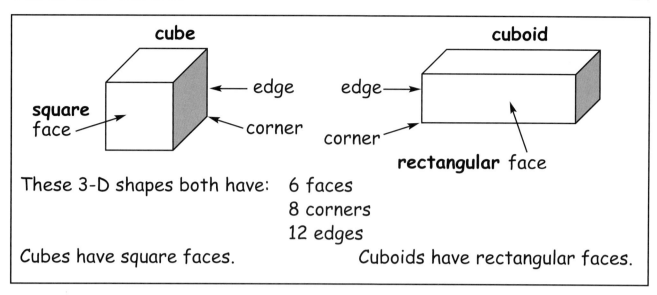

These 3-D shapes both have: 6 faces
8 corners
12 edges

Cubes have square faces. Cuboids have rectangular faces.

1 Write the name of the 3-D shape under each object.

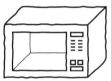    

a _____ b _____ c _____ d _____

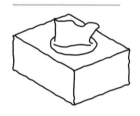   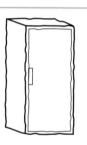

e _____ f _____ g _____ h _____

2 This is not a cuboid.

   a Why not? _____

   b What shape is it? _____

3 This is not a cube.

   a Why not? _____

   b What shape is it? _____

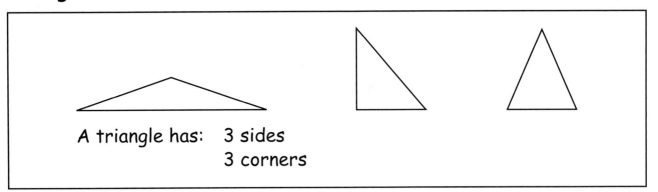

A triangle has: 3 sides
3 corners

1 Tick the triangles.

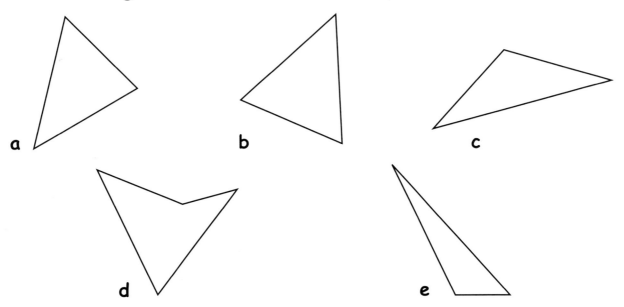

a

b

c

d

e

2 How many triangles can you see in this butterfly? _____

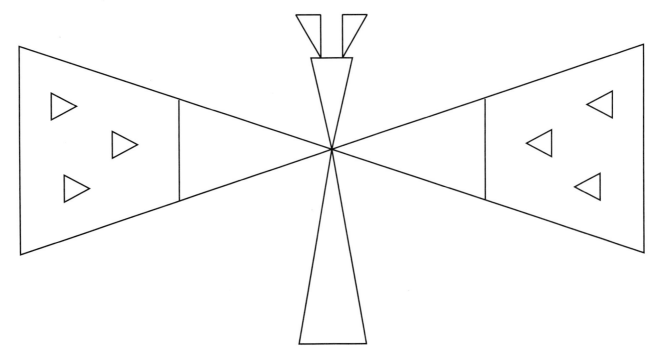

2-D shapes

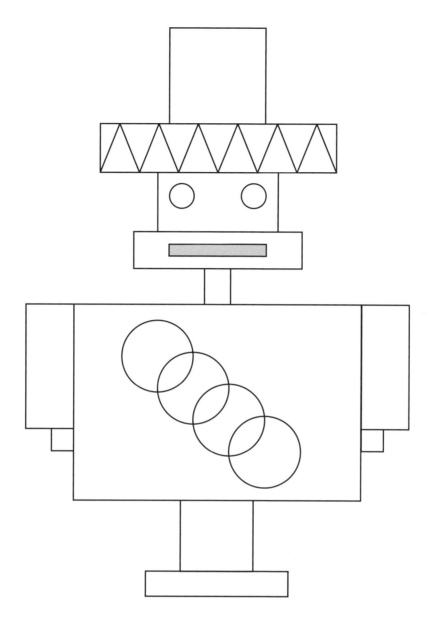

1 How many rectangles can you see? _____

2 How many squares can you see? _____

3 How many circles can you see? _____

4 How many triangles can you see? _____

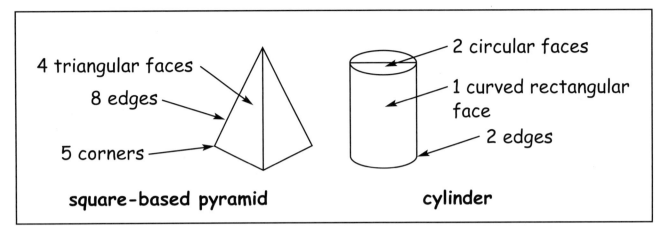

4 triangular faces

8 edges

5 corners

square-based pyramid

2 circular faces

1 curved rectangular face

2 edges

cylinder

1 Write the name of the 3-D shape for each item.

a _____

b _____

c _____

d _____

e _____

f _____

2 Fill in the missing words.

circular	triangular	rectangular
square	edges	corners	two

a A cylinder has _____ edges and two _____ faces.

It has one curved _____ face.

b A square-based pyramid has a _____ base and four

_____ faces. It has eight _____ and five

_____ .

Sides and corners

1 Draw lines to match the shapes, names and number of sides
   and corners.

shape	name	sides	corners
a	triangle	4 sides – usually 2 equal long sides and 2 equal short sides	4 corners
b	rectangle	4 equal sides	3 corners
c	square	3 sides	4 corners

2 Which shape has 3 sides? _____

3 Which shape usually has 2 long sides and 2 short sides? _____

4 Which shapes have 4 corners? _____ and _____

5 How is a square different from a rectangle? _____

6 When is a rectangle a square? _____

2-D faces on 3-D shapes   E2

1 Draw lines to join each 3-D shape to its 2-D faces.

Remember Some 3-D shapes have faces that are different shapes,
so you may need more than one line from each 3-D shape.

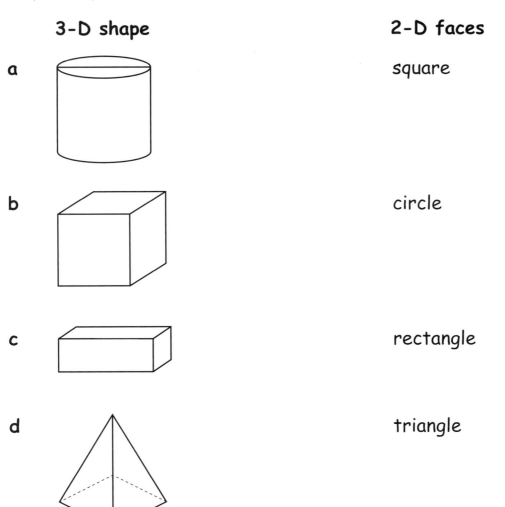

3-D shape	2-D faces
a	square
b	circle
c	rectangle
d	triangle

2 Tick the boxes to show the shape of each 3-D shape's 2-D faces.

3-D shape	shape of 2-D face			
	square	rectangular	circular	triangular
cube				
cuboid				
cylinder				
square-based pyramid				

Shape and space

Faces, edges and corners

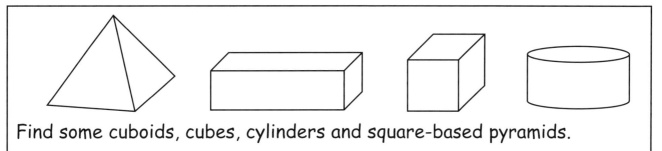

Find some cuboids, cubes, cylinders and square-based pyramids.

For each shape count the: number of faces
                         number of edges
                         number of corners

1 Fill in the missing words.

six	twelve	eight	eight	
five	five	three	two	no

   a Cubes and cuboids have _____ faces, _____ edges and

      _____ corners.

   b Cylinders have _____ faces, _____ edges and _____
      corners.

   c Square-based pyramids have _____ faces, _____ edges and

      _____ corners.

2 a Is a cube the same as a cuboid? _____

   b Why? _____

3 Why do you only need one measurement for a cube?

   _____

4 a How many measurements do you need for a cuboid? _____

   b What are they? _____

Where's the shape?

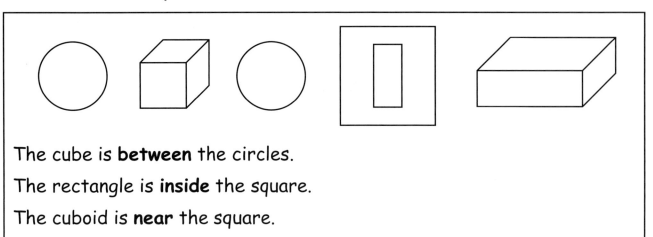

The cube is **between** the circles.

The rectangle is **inside** the square.

The cuboid is **near** the square.

1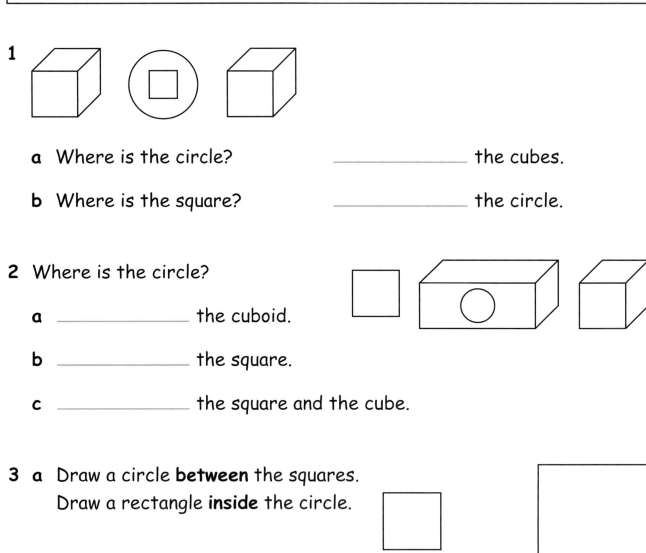

   a Where is the circle?   _____ the cubes.

   b Where is the square?   _____ the circle.

2 Where is the circle?

   a _____ the cuboid.

   b _____ the square.

   c _____ the square and the cube.

3 **a** Draw a circle **between** the squares.
       Draw a rectangle **inside** the circle.

    b Draw a circle **near** the rectangle.
       Draw a square **near** the cuboid.

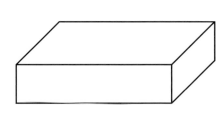

Shape and space

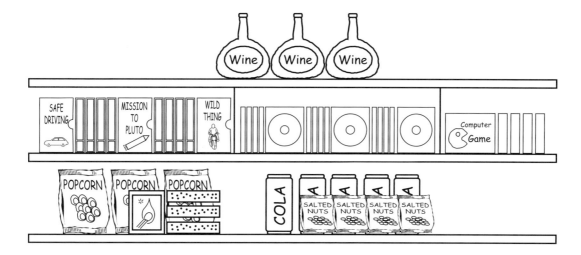

The videos are **to the left** of the DVDs.
The computer games are **to the right** of the DVDs.
The wine is **above** the DVDs.
The cans of cola and nuts are **below** the DVDs.
The cans of cola are **behind** the nuts.
The nuts are **in front** of the cans of cola.

1 What is to the **right** of the DVDs? _____

2 What is to the **left** of the computer games? _____

3 What is **above** the cans of cola? _____

4 What is **below** the wine? _____

5 What is **behind** the matches? _____

6 What is **in front** of the pop corn? _____

7 These shelves need filling.
  Write bread to the **left** of the biscuits.
  Write sugar to the **right** of the biscuits.
  Write coffee **above** the biscuits.
  Write tea **below** the biscuits.
  Draw a bag of rice **behind** the biscuits.
  Draw a triangle **in front** of the biscuits.

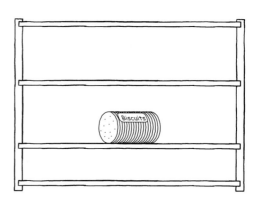

Shape and space skill check 1 E1

p71-75 ▶

1 Write the name of each shape.

| circle | rectangle | square |
| cube | cuboid | |

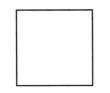

a _____ b _____ c _____

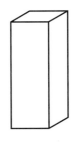

d _____ e _____

2 Tick the 2-D shapes.

    circle   cube   cuboid   square   rectangle

p82 ▶

3 a Draw a square between the rectangles.

  b Draw a circle inside the square.

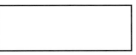

4

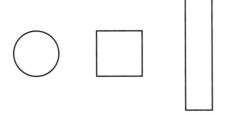

| near | between | inside |

Fill in the missing words.

  a The circle is _____ the square.

  b The square is _____ the circle and the rectangle.

Shape and space skill check 2

p76-78

1 Write the names of the shapes.

circle	square	cylinder
rectangle	triangle	cube cuboid

    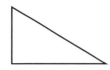 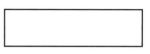

a _____ **b** _____ **c** _____ **d** _____

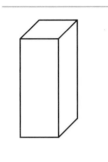

e _____ **f** _____ **g** _____

p78-81

2 Complete the table by filling in the correct number of faces, edges and corners for each shape.

3-D shapes	number of faces	number of edges	number of corners
cube			
cuboid			
cylinder			
square-based pyramid			

p83

3 a Draw a circle to the left of the square.

   b Draw a triangle to the right of the square.

   c Draw a rectangle above the square.

   d Write an X below the square.

Skill check

Shape and space skill check 1 answers

1 a square   **b** circle   **c** rectangle   **d** cuboid   **e** cube

2 shapes ticked: circle, square, rectangle

3 a b

4 a near   **b** between

Shape and space skill check 2 answers

1 a square   **b** circle   **c** triangle   **d** rectangle

    e cuboid   **f** cube   **g** cylinder

2

3-D shapes	number of faces	number of edges	number of corners
cube	6	12	8
cuboid	6	12	8
cylinder	3	2	0
square-based pyramid	5	8	5

3 a b c d

1 The Nature Centre opens at 10 o'clock.

Your watch says:

Is the Nature Centre open? _____

2 It costs £2 to go to the Nature Centre.

Which coin will you use? _____

    a   b   c   d   e   f

Feeding times	
rabbits	11 o'clock
otters	1 o'clock
pigs	2 o'clock
deer	3 o'clock

3 a Which animals are fed in the morning? _____

   b Which animals are fed in the afternoon? _____

4 The keeper records the animals' food each day.
Write the days of the week in order starting with Monday.

Tuesday   Saturday   **Monday**

       Sunday   _____

             Friday   _____

Thursday   Wednesday   _____

                                     _____

                                     _____

Mock test

deer   rabbit   pony

5 a Which animal is the largest? _____

b Which animal is the smallest? _____

6 Fill in the missing words.

a The deer is _____ than the pony.

b The deer is _____ than the pony.

c The pony has a _____ tail.

d The pony is _____ than the rabbit.

taller	shorter

wider	thinner

long	short	wide

lighter	heavier

7 Which animal is the lightest ? _____

8 You have 3 drinks.

a Which glass holds the most? _____

b Which glass holds the least? _____

c Which glass is empty? _____

9 Name the shape of these signs.

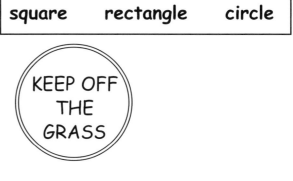
square   rectangle   circle

a _____

b _____

c _____

10 Name the shape of these animal homes.

cube   cuboid

a _____

b _____

11

Fill in the missing words:

near   between   inside

a The cat is _____ the dog.

b The fish is _____ the bowl.

c The dog is _____ the cat and the fish.

Mock test

1 Pat is invited to a wedding on 8th December 2005.
    Which of the following say the same date?
    Circle the answer.

08/12/04   08/12/05   12/12/05

2 The wedding is at quarter past three.
    Write this time on the digital clock.

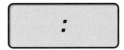

3 Pat buys a card for 95p.
    Which set of coins will pay for the card?

a
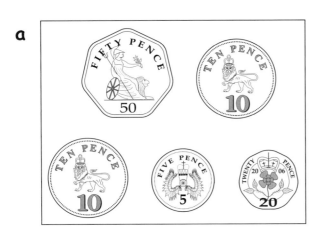

b
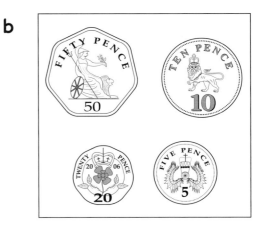

4 Pat buys paper to wrap the present for 45p and a bow for 26p.

   a How much is this altogether? _____

   b How much change will she have from £1? _____

5 Pat makes a tablecloth for a wedding present

   a The tablecloth needs to be 1 m long.
       She has a piece of material 95 cm long.
       Is this long enough? _____

b Pat draws a pattern to sew on the tablecloth.
Which units are used to measure the pattern?
Circle the answer.

cm   m   ml   kg

c Use a ruler to find the length of the pattern. _____

 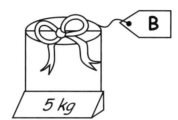 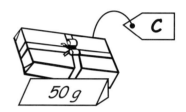

6 a Which present is the heaviest? _____

b Which present is the lightest? _____

c Which present is most likely to be Pat's tablecloth? _____

7 What is the weight of

present D? _____

8 What is **kg** short for? Circle the answer.

centimetres   kilograms   grams   millilitres

9 How much wine is in the jug?

_____

Measures, shape and space mock tests

10 What is **l** short for? Circle the answer.

millilitre   **centimetres**   **litre**   **gram**

11 What temperature is shown on the thermometer? _____

°Celsius

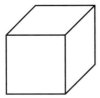 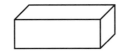 

12 a Draw a triangle around the cylinder.

  b Draw a circle around the square-based pyramid.

13 a How many faces does a cube have? _____

  b How many faces are there on a cylinder? _____

  c How many edges are there on a square-based pyramid? _____

  d How many corners are there on a cube? _____

  e Which 3-D shape has circular and rectangular faces? _____

  f Why do you only need one measurement for a cube?

_____

Mock test

Money

Page 1

Choose coins
1 50p
2 20p
3 £1
4 10p

Page 2

Which coin?
1
2
3
4
5
6
7
8
9 1p, 2p, 5p, 10p,
   20p, 50p, £1, £2

Page 3

Choose the £s
1
2
3
4
5 £10, £5, £2, £1

Page 4

Up to 10p
1 5p
2 7p
3 7p
4 8p
5 7p
6 8p
7 8p
8 4p
9 7p
10 9p

Page 5

Up to 20p
1 13p = b and f
   14p = c and d
   15p = a and e
2 a, e, f should be
   ticked

Page 6

Up to 50p
1 a, b, d
2 a 10p
   b 20p
   c 50p
   d 50p
   e 20p

Page 7

Which coins?
1 20p = c and f
   30p = b and d
   40p = a and e
2 a, d, e

Page 8

Find the coins
1 a 20p, 5p, 2p
   b 50p, 10p, 5p
   c 10p, 2p, 1p
   d 10p, 5p, 2p
   e 50p, 20p, 10p
   f 10p, 2p, 2p
   g 50p, 20p, 5p
2 a 10p, 5p
   b 5p, 2p
   c 50p, 5p
   d 20p, 10p
   e 20p, 5p

Page 9

Change from £1
1 total 78p
   change 22p
2 total 89p
   change 11p
3 total 97p
   change 3p
4 total 77p
   change 23p
5 total 54p
   change 46p
6 total 79p
   change 21p
7 total 77p
   change 23p
8 total 96p
   change 4p
9 total 88p
   change 12p

Page 10

Shopping problems
1 26p
2 4p
3 6p
4 5p
5 1p
6 a 88p   b 12p
7 a 99p   b 1p
8 a 68p   b 32p
9 a 84p   b 16p

Page 11

Shopping with £1
1 total 71p
   change 29p
2 total 93p
   change 7p
3 total 52p
   change 48p
4 total 56p
   change 44p
5 total 72p
   change 28p
6 total 73p
   change 27p
7 total 91p
   change 9p
8 total 94p
   change 6p
9 total 76p
   change 24p

Page 12

More shopping problems
1 17p
2 6p
3 8p
4 19p
5 a 66p
   b 34p
6 a 36p
   b 14p
7 a 87p
   b 13p
8 a 96p
   b 4p
9 a 63p
   b 37p

Page 13

Calculate with whole £s

1. Mark   total £16 change £4
   Paul   total £15 change £5
   Soumia total £17 change £3
2. £10   **3** £5   **4** no they only have £1

5.

meal	cost	money left over			
		Deb £10	Max £12	Yui £15	Jag £20
kebab and chips	£4	£6	£8	£11	£16
curry and chips	£6	£4	£6	£9	£14
2 burgers and 1 chips	£5	£5	£7	£10	£15
pizza, kebab and chips	£8	£2	£4	£7	£12

Time

Page 17

Days of the week

1. Sunday, Monday, Tuesday, Wednesday, Thursday, Friday, Saturday
2. Tuesday
3. Friday
4. Wednesday
5. Friday
6. Saturday
7. Saturday and Sunday
8. Friday and Sunday

Page 19

Write the date

1. a 4/Nov/2005   b 1/Jan/2006
   c 23/Mar/1995   d 2/Oct/1963
2. a 21/08/01   b 12/05/03
   c 3/11/05   d 11/06/99
3. a 16th September 2004
   b 29th April 2005
   c 9th December 2002
   d 15th October 2005
4. 31/08/06

Page 18

Seasons

1. a summer
   b winter
   c autumn
   d spring
2. spring
3. autumn
4. winter
5. winter
6. summer
7. autumn
8. spring, summer, autumn, winter

Page 20

Match the dates

7th January 2006	4 a
13th March 2006	7 c
18th August 2006	2 m
12th June 2006	1 k
1st November 2006	6 l
30th September 2006	14 j
1st April 2006	8 b
2nd December 2006	9 g
3rd July 2006	3 f
5th February 2006	11 i
21st June 2006	5 d
14th May 2006	13 h
22nd October 2006	12 e
31st January 2006	10 n

Page 21

Match the times

1. b   **2** a   **3** d   **4** c
5. a 10 o'clock   b 7 o'clock   c 1 o'clock
   d 3 o'clock
6. a
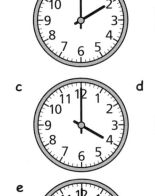
b

c
d

e

Page 23

Vet's weekly plan

days	morning am		afternoon pm	evening pm	
Sunday					
Monday	Sam		Bob	Molly Sandy	
Tuesday	Rover		Gypsy Rex Daisy		
Wednesday					
Thursday			Jodie Pippa		
Friday	Milo				
Saturday	Ben				

2 Sunday and Wednesday

3 Tuesday

Page 24

Hours

1 **a** 3 o'clock   **b** 10 o'clock   **c** 1 o'clock   **d** 6 o'clock
  e 11 o'clock **f** 9 o'clock   **g** 7 o'clock   **h** 4 o'clock

2 **a**   **b**   **c**    **d**

3 **a** 12:00   **b** 10:00   **c** 07:00   **d** 01:00

Page 25

Half hours

1 **a** half past 7   7 thirty   7:30   **b** half past 9   9 thirty   9:30
  c half past 4   4 thirty   4:30   **d** half past 10 10 thirty 10:30
  e half past 11 11 thirty 11:30

2 **a**   **b**    **c**    **d**

3 **a** 09:30   **b** 11:30   **c** 03:30   **d** 07:30

Page 26

Match the times again

1 twelve thirty   `12:30`   2 half past three   `03:30`

3 half past one   `01:30`   4 nine o'clock   `09:00`

5 nine thirty   `09:30`   6 half past eight   `08:30`

7 six o'clock   `06:00`   8 ten thirty   `10:30`

Page 27

Quarter hours

1 a quarter past 11   11 fifteen   11:15   b quarter past 6   6 fifteen   6:15
   c quarter past 4   4 fifteen   4:15   d quarter past 7   7 fifteen   7:15
   e quarter past 9   9 fifteen   9:15

2 a    b   c   d

3 a `08:15`   b `09:15`   c `05:15`   d `01:15`

Page 28

Quarter to

1 a quarter to 10   9 forty-five   9:45   b quarter to 2   1 forty-five   1:45
   c quarter to 4   three forty-five   3:45   d quarter to 7   six forty-five 6:45
   e quarter to 9   eight forty-five   8:45

2 a `07:45`   b `05:45`   c `02:45`   d `05:45`

Page 29

Quarter past or quarter to?

1 a quarter past 7   b quarter to 8   c quarter past 9   d quarter to 11
   e quarter past 11   f quarter to 3   g quarter past 3   h quarter to 4

2 a    b    c    d

   e   f    g    h

Page 30

Match more times

1 quarter past 12   `12:15`

2 quarter past six   `06:15`

3 quarter to ten   `09:45`

4 quarter to eight   `07:45`

5 quarter past nine   `09:15`

6 quarter to nine   `08:45`

7 quarter to three   `02:45`

8 quarter past two   `02:15`

Page 31

Find the time

1 yes

2 half past 4 or 4:30

3 `09:15`

4 yes

5 no

6 half past 1 or 1:30

7 `07:45`

Answers

Measures

Large or small?
1 a small
  b large
  c small
  d large
  e small
  f large
2 a smaller
  b larger
  c smaller
  d larger
3 a largest A
    smallest B
  b largest B
    smallest A

Long, wide, tall, short, thin
1 a long, thin, short
  b tall, thin
  c short, wide
  d short, thin
2 a longer
  b shorter
  c wider
  d thinner
3 a 1 and 2
  b 1
  c 3
  d 2 and 3

Longest, shortest, widest, narrowest, thinnest
1 a B
  b A
  c C
2 a A
  b C
  c C
3 a C
  b A
  c B
4 a longest
  b shortest
  c thinnest
5 a widest
  b longest
  c shortest

Measure in centimetres
1 9 cm
2 5 cm
3 3 cm
4 8 cm
5 2 cm
6 11 cm
7 10 cm
8 6 cm
9 5 cm
10 12 cm
11 14 cm
12 15 cm
13 9 cm
14 14 cm
15 16 cm

More measurements
1 7 cm
2 3 cm
3 3 cm
4 19 cm
5 8 cm
6 5 cm
7 14 cm
8 5 cm
9 6 cm
10 7 cm
11 4 cm
12 3 cm
13 a 30 cm
   b 21 cm

Measure the sides
1 a 6 cm
  b 3 cm
2 5 cm
3 a 8 cm
  b 3 cm
4 4 cm
5 a 11 cm
  b 3 cm
  c 10 cm
6 a 5 cm
  b 2 cm

Estimate and measure
Ask your maths tutor to check your work.

Metres and centimetres
1 a cm   b m
  c cm   d m
  e m   f m
  g cm   h cm
  i cm   j m
2 a no   b yes
  c no   d yes
  e yes   f no
3 a 120 cm
  b 110 cm
  c 140 cm
  d 130 cm

Heavy or light?
1 a heavy
  b light
  c heavy
  d light
  e light
  f light
  g heavy
2 a heavier
  b lighter
  c heavier
  d lighter
  e heavier
  f heavier
  g lighter
  h heavier
  i lighter
  j lighter
  k lighter

Heaviest or lightest?
1 a pan   b coin
2 a balloon
  b brick
3 a computer
  b feather
4 a heaviest
  b lightest

Kilograms or grams?
1 a g
  b g or kg
  c kg   d g
  e kg
  f kg
2 a 1 kg   b 2 kg
  c 100 g   d 50 g

More or less than 1 kilogram?
Items ticked are: a, b, c, d, f, h, j, l, m, n, p

Estimate and weigh
Ask your maths tutor to check your work.

More or less than $\frac{1}{2}$ kilogram?
1 a, b, e, f
2 a 1 kg
  b more than 3 kg
  c about $1\frac{1}{2}$ kg
  d about $\frac{1}{2}$ kg
  e 2 kg

Reading scales
1 B 5 kg   C 9 kg   D 12 kg   E 16 kg
  F 20 kg   G 23 kg   H 31 kg   I 38 kg
2

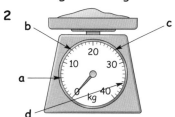

Answers

Page 50

Empty, full, more or less?
1 a A   b B
2 a B   b A
3 a empty
   b more
4 a full
   b more
5 a full
   b less
   c less

Page 51

Which holds most?
1 a 1
   b the same height but wider
2 a 1
   b the same height but wider
3 a 2
   b same width but taller
4 a 2
   b same height but wider

Page 52

Litres and millilitres
1 l   2 ml
3 l   4 ml
5 l   6 ml
7 l   8 ml
9 a 2 l
   b 500 ml
   c 5 ml
   d 200 ml
   e 1 l

Page 53

More or less than 1 litre?
1 items ticked: b, c, d
2 a 4   b 5
   c 7   d 10
3 items ticked: a, b

Page 54

More or less than ½ litre?
1 items ticked: b, d, f, i
2 a ½ litre
   b 1½ litres
   c 1 litre
   d 2½ litres
   e 2 litres

Page 55

Measure in litres
1 2 l
2 1½ litres, 1500 ml or 1 l 500 ml
3 500 ml or ½ litre
4 3 l   5 4 l   6 5 l
7 3½ litres, 3500 ml or 3 l 500 ml
8 4½ litres, 4500 ml or 4 l 500 ml

Page 56

Hot or cold?
1 pictures with circles: c, d, g
2 pictures with squares: b, h, i
3 pictures ticked a, e, f

Page 57

Warmer or cooler?
1 cooler
2 warmer
3 warmer
4 cooler
5 warmer
6 cooler
7 cooler
8 warmer
9 cooler
10 warmer

Page 58

What is the temperature?
Dublin 13°C
London 14°C
Bonn 18°C
Berlin 16°C
Berne 22°C
Vienna 21°C

Page 59

Weather forecasts
1 Bangkok, Colombo, Dakar, Khartoum, Singapore
2 Auckland, Baghdad, Hong Kong, Islamabad, London, Melbourne, New Orleans, Quinto, Rome, Tokyo
3 Berlin, Budapest, Edinburgh, Kiev, Ottawa

Page 60

Read a thermometer
1 2 4
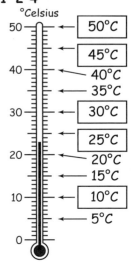
3 10°C

Page 61

Scales around the house
1 a 4   b 3
   c 2   d 1
2 a 1   b 4   c 3
3 a 100 b 150 c 200
4 a 2   b 4   c 5

Page 62

Dials in the car
1 a 21 or 22 mph
   b 50 mph
   c 45 mph
2 a more
   b more
   c less
3 a 27, 28 or 29
   b 6, 7 or 8
   c 21 or 22

Page 63

Fill in the gaps

1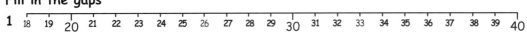

2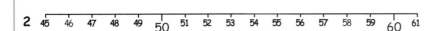

3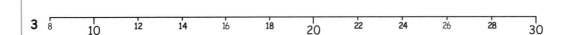

4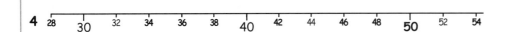

5 a 16   b 24

Answers

Page 64

More gaps

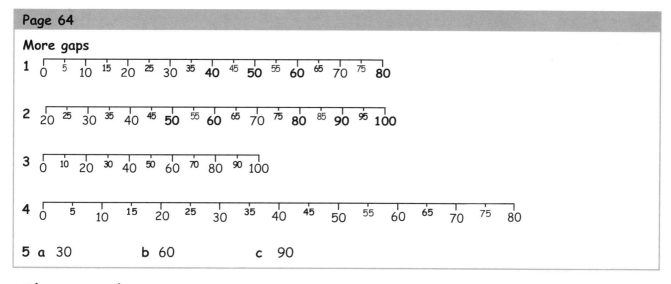

1 0 5 10 15 20 25 30 35 **40** 45 **50** 55 **60** 65 70 75 **80**

2 20 25 30 35 40 45 **50** 55 **60** 65 70 75 **80** 85 **90** 95 **100**

3 0 10 20 30 40 50 60 70 80 90 100

4 0 5 10 15 20 25 30 35 40 45 50 55 60 65 70 75 80

5 a 30 b 60 c 90

Shape and space

Page 71

Squares
1 b and f 2 18
3 Ask your maths
  tutor to check
  your answer.

Page 72

Rectangles
1 a and e
2
a rectangle b square
c rectangle d square
3 a 17 rectangles
  b 8 squares

Page 73

Circles
1 a and d 2 10
3 Ask your maths
  tutor to check
  your answer.

Page 74

Which shape?
1 a

square

b

rectangle

c

circle

2
a circle b rectangle
c square d circle
e square f rectangle

Page 75

Cubes and cuboids
1 a cuboid
  b cube
  c cuboid
  d cube
  e cuboid
  f cube
  g cuboid
  h cuboid
2 a each face is a
    square
  b cube
3 a some faces are
    rectangular
  b cuboid

Page 76

Triangles
1 a, b, c, e
2 14

Page 77

2-D shapes
1 9
2 4
3 6
4 13

Page 78

Pyramids and cylinders
1 a cylinder   b cuboid   c cylinder
  d square-based pyramid   e cube
  f square-based pyramid
2 a A cylinder has **two** edges and two **circular**
    faces. It has one curved **rectangular** face.
  b A square-based pyramid has a **square** base
    and four **triangular** faces. It has eight
    edges and five **corners**.

Page 79

Sides and corners
1 a square   4 equal sides   4 corners
  b triangle   3 sides   3 corners
  c rectangle   4 sides, usually 2 equal long
                sides and 2 equal short sides 4 corners
2 triangle   3 rectangle
4 square and rectangle
5 all the sides are equal
6 when all the sides are equal

Page 80

2-D faces on 3-D shapes
1 a circle, rectangle   b square
  c rectangle, square   d square, triangle
2

3-D shape	shape of 2-D face			
	square	rectangular	circular	triangular
cube	/			
cuboid	/	/all or some		
cylinder		/	/	
square-based pyramid	/			/

Answers

Page 81

Faces, edges and corners

1 a Cubes and cuboids have **six** faces, **twelve** edges and **eight** corners.

b Cylinders have **three** faces, **two** edges and **no** corners.

c Square-based pyramids have **five** faces, **eight** edges and **five** corners.

2 a no

b a cube has all square faces

3 all the sides are equal on a cube

4 a 3

b length, width and height

Page 82

Where's the shape?

1 a between the cubes

b inside the circle

2 a inside the cuboid

b near the square

c between the square and the cube

3 a

b Ask your maths tutor to check your answer.

Page 83

On the shelves

1 computer games

2 DVDs

3 DVDs

4 DVDs

5 pop corn

6 matches

7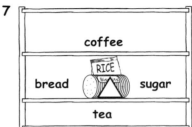

Measures, shape and space mock tests

Page 87-89

Measures, shape and space mock test 1

1 no

2 d

3 a rabbits   b otters, pigs, deer

4 Monday
Tuesday
Wednesday
Thursday
Friday
Saturday
Sunday

5 a deer   b rabbit

6 a taller   b thinner
c long   d heavier

7 rabbit

8 a A   b B   c C

9 a rectangle   b circle   c square

10 a cuboid   b cube

11 a near   b inside   c between

Page 90-92

Measures, shape and space mock test 2

1 08/12/05

2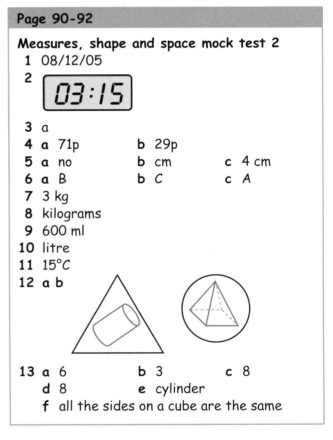

3 a

4 a 71p   b 29p

5 a no   b cm   c 4 cm

6 a B   b C   c A

7 3 kg

8 kilograms

9 600 ml

10 litre

11 15°C

12 a b

13 a 6   b 3   c 8
d 8   e cylinder
f all the sides on a cube are the same